Diego Marloni dos Santos

Energy Management in Industries

Diego Marloni dos Santos

Energy Management in Industries

How do the industries of Ponta Grossa - PR manage their energy resources?

ScienciaScripts

Imprint

Cover image: www.ingimage.com

This book is a translation from the original published under ISBN 978-613-9-65855-8.

Publisher:
Sciencia Scripts
is a trademark of
Dodo Books Indian Ocean Ltd. and OmniScriptum S.R.L publishing group

120 High Road, East Finchley, London, N2 9ED, United Kingdom
Str. Armeneasca 28/1, office 1, Chisinau MD-2012, Republic of Moldova, Europe
Printed at: see last page
ISBN: 978-620-7-97137-4

I dedicate this work to my grandparents, Tereza Roseli dos Santos and João Ataídes dos Santos, because of my great admiration and respect for them, as without them this work and many of my dreams would not come true.

ACKNOWLEDGEMENTS

To God for giving me the health and strength to overcome difficulties, and especially for giving me the dedication and persistence to complete this work.

To my paternal grandmother, Tereza Roseli dos Santos, for being my first and eternal teacher, not just for teaching me how to read and write, but for giving me the tools to be a man of character and a dreamer today. Her teachings will always be at the forefront of my goals.

To my paternal grandfather, João Ataídes dos Santos, for having been much more than a grandfather, a father for all hours, always showing love and affection, passing on confidence and motivation through his teachings, which will always carry my admiration.

SanfAna College, its teaching staff, management and administration, which provided the window through which I can now glimpse a higher horizon, especially the Logistics Technology course, which opened its doors to the realisation of a dream.

To the coordinator of the Logistics Technology course, Professor Balduir Carletto, for his extreme encouragement, support, teaching, dedication and confidence in my potential, in which he has been doing an excellent job in the administration of the course, receiving many compliments from students and teachers.

To my supervisor, Professor Isabel Márcia Rodrigues, for her support in the short time she had, for her corrections, encouragement and, above all, for motivating me through her teachings to carry out this work, always believing in my ability. This motivation was fundamental to the realisation of this Course Conclusion.

To my classmates who, since the beginning of the course, have stuck together, both in good times and in times of difficulty. They were always there to support me in my academic endeavours.

To everyone who directly or indirectly played a part in this achievement, thank you very much.

SUMMARY

The aim of this Final Year Programme was to investigate how industries in the Ponta Grossa region deal with energy management. It was based on the Brazilian energy scenario, with the industrial sector being responsible for the consumption of almost half of the electricity supplied in the country, which has now become one of the most essential consumer goods for society. In order to achieve the objective proposed in this research, the deductive method was used, followed by descriptive research. With regard to the problem, a quantitative approach was used. As a research technique, the procedure adopted was a survey targeting the industries of Ponta Grossa, where a questionnaire was applied to 10 industries. The questionnaire contained 15 objective questions and 2 descriptive questions to question people from the companies chosen, in order to collect data and information relevant to analysing and concluding the scenario under study. With regard to the theoretical basis, the themes of energy efficiency, energy and the environment and the industrial scenario were researched. It is believed that the results of this research will help to raise awareness of the importance of energy issues, and that industries in other locations will be able to recognise the importance of managing this resource and adopt some of the practices described in the study. The result of this research also aims to contribute data for further studies related to the city of Ponta Grossa, which is now considered one of Paraná's industrial centres.

Keywords: Energy management. Evaluation. Industrial scenario.

SUMMARY

CHAPTER 1

INTRODUCTION

The availability of electricity is essential for a wide range of activities, the main ones being industry, transport and commercial and residential consumption (FARIA, 2011). A determining factor for the country's economic development and also for human actions is society's access to mechanical, thermal and electrical energy infrastructure services (ANEEL, 2007).

Brazil has made great strides in this area, and its energy production is largely concentrated in hydroelectric power stations, which, with the huge growth in population and industry, has greatly increased the country's consumption of this commodity. The large economy is driving the increase in consumption, household appliances are becoming more and more accessible and technology is constantly increasing the need for energy to function.

1.1 Presentation of the research topic

Given this context, the topic addressed in this work is energy management in the industries of Ponta Grossa, seeking to investigate how they manage this resource.

1.2 Presentation of the research problem

Globalisation has made companies increasingly competitive, so strategy and technology are basic requirements for many to remain in the market. It can be seen that good management should be one of the objectives of any well-run industry. A company with efficient energy management can achieve significant results such as reducing electricity costs and collaborating with the preservation of natural resources.

In view of this discussion, the problem question of this research is: How do industries in the city of Ponta Grossa manage their energy resources?

1.3 Justification

Energy is one of the fundamental elements of life and the driving force behind progress, because unlike animals, in order to facilitate their habitat and life, such as food, comfort and luxury goods, among others, humans have an impact on the environment in which they live, thus depleting natural resources. Energy is one of the main elements for the development of humanity, which occupies a privileged space in contemporary society through its various forms and purposes, passed on through the burning of fossil fuels, nuclear, hydroelectric and thermoelectric plants, energies that are used by society as a whole (FRAGOMENI and GOELLNER, 2009).

The subject of energy has once again become a priority in various discussions and world conferences, for different reasons: pre-salt, natural gas, renewable energies, biodiesel, economic geopolitics, the greenhouse effect and, above all, environmental issues (FRAGOMENI and GOELLNER, 2009).

The focus of this study is on electricity, which has now become one of the most essential consumer goods for modern society. At present, everything needs energy to function, such as the generation of lighting, the movement of machinery and equipment, communication, temperature control, among others. Quality of life depends on energy to function, among other factors such as mobility, transport, efficiency, safety and comfort (MMA, 2012).

The region of Ponta Grossa has recently stood out due to the large migration of companies and also because it is one of the largest industrial centres in the state. As such, the increase in electricity consumption in the region is evident, which makes it necessary to assess how the city has been progressing in energy management and what impacts this sector may have in the future.

It is believed that the results of this research will help to promote the importance of raising awareness of energy issues, as well as publicising the possibility of organisations consuming electricity in a sustainable and rational way, helping to reduce the consumption of natural goods and environmental impacts.

1.4 Objectives

1.4.1 General Objective

To investigate how industries in Ponta Grossa act on the issue of energy management.

1.4.2 Specific objectives

- Identify whether companies have an energy management policy.
- Present the sources of energy consumption in industry.
- Identify the energy efficiency actions that organisations use.

1.5 Organising the study

This work is organised as follows:

Chapter 1 - Introductory part, followed by the justification, presentation of the topic, the problem and the research objectives.

Chapter 2 - Theoretical Framework: This begins with the relationship between logistics and energy, followed by the analogy between energy and the environment, since both are fundamental to good energy management, where it is necessary to emphasise the importance of environmental issues in these sectors. The research also continues with a theoretical framework, starting from a broad scenario to a more specific one, describing the international energy panorama, the energy scenario in Brazil and finally the industries.

Chapter 3 - Methodology: presents the methodological part of the research, starting with the methods adopted and their classifications, which are tools that validate this research. It also seeks to explain the reasons and techniques used to choose the universe under analysis and to collect and analyse the data.

Chapter 4 - Presentation and interpretation of data and results: The data collected in the research is presented, analysed and discussed.

Chapter 5 - Conclusions: This chapter presents the final considerations about the study, as well as the results obtained and suggestions for future studies and analyses in this sector.

CHAPTER 2

THEORETICAL BACKGROUND

2.1 Logistics and Energy

In recent years, society has been demanding that organisations improve their business, economic and social processes, especially in relation to environmental issues. Aware of this discussion, Gomes and Tortato (2010) report that logistics has aroused great interest within organisations, in which scientific research can generate competitive advantages for companies, as well as contributing to the issue of corporate and environmental sustainability.

Energy efficiency is an activity that seeks to optimise the use of energy sources. In this way, it consists of using less energy to provide the same amount of energy value needed to run equipment, machines, processes and so on (GOMES and TORTATO, 2010).

Ballou (2006) defines the term logistics as a branch of military science that deals with the acquisition, maintenance and transport of material, personnel and installations.

The concept of logistics has become specialised in the efficiency of tools and management of the return flow of materials and products, due to the imposition of current market needs together with the emergence of new technologies in the area of reverse logistics. In addition, the competitiveness of companies in line with customer demands and government legislation has encouraged increased efficiency in reverse logistics (GOMES and TORTATO, 2010).

One determining factor that impacts a company's fixed costs is electricity costs, which often determine the success or failure of an organisation. Companies that plan properly in this regard can significantly reduce their electricity costs (MAYER; CASTELLANELLI; HOFFMANN, 2007).

Lately, logistics has been able to position itself and get used to this global trend, mainly in the search for high efficiency conditions, also in the coordination of the various links in the supply chain and in reducing lead times and operating costs. However, in order to acquire a differentiated stance in the energy sector, logistics operators will have to offer innovative resources in order to add value to their customers' supply chain, reducing costs and resulting in a higher quality of service and efficiency (LARANJEIRA, 2015).

The relationship between logistics and energy occurs in various ways. One of these is the fact that logistics processes are involved in the transport of natural gas, which provides energy for industries via gas pipelines, while water and electricity are also transported. In addition, other products can be transported by pipeline, such as crude oil and its derivatives (BARROS et al. 2008). Another issue is the installation of electricity circuits which, through aerial networks supported by poles or underground pipelines, distribute energy that is delivered to consumers (PORTAL BRASIL, 2014).

2.2 Energy and the Environment

Energy is a fundamental ingredient for human life. At the time of primitive society, its cost was practically zero. Energy was obtained from forest firewood for domestic use in the kitchen, and as a source of light and heat. Gradually, however, energy consumption grew more and more, leading to the need for other energy sources. During the Middle Ages, water and wind sources were used in insufficient quantities to supply the growing population, especially in cities. Other types of energy with high production costs, such as coal, oil and gas, were used after the Industrial Revolution to meet the demands of society in general (GOLDEMBERG and LUCON, 2007).

The production and commercialisation of energy are issues of economic, political and social development for any country, due to the relationship between man's contemporary dependence on this input. The great competitiveness between nations and the search for efficient energy solutions, which cause less environmental impact and bring economic and social benefits, has increasingly driven the implementation of Research and Development activities in the Energy Sector (NASCIMENTO et al, 2013).

As an activity that has a significant environmental impact, the energy sector is currently under intense discussion around the world, especially in relation to sustainability and human activities. The exploration, transport and processing of non-renewable energy resources, such as fossil fuels, has a major environmental impact and also harms human health. A major determining factor of these fuels is the release of greenhouse gases through their use and processing, which is an important input for electricity generation in many countries (GOLDEMBERG and LUCON, 2007).

One of the most complex and discussed issues in the world today is the effects of CO2 gas emissions in the atmosphere. Emissions from the energy sector are one of the biggest global problems related to climate change. These gases, which accumulate in the atmosphere, cause the greenhouse effect to the point of disrupting the climate patterns that condition the life of living beings, in which the growing concentration of CO2 in the atmosphere and the increase in average temperatures are becoming increasingly evident. This is having a major impact on agricultural productivity, fishing, flooding of coastal regions and an increase in natural disasters (JANNUZZI, 2001).

Also according to Jannuzzi (2001), the production of electricity in thermoelectric plants accounts for around a third of carbon dioxide emissions worldwide, followed by the transport and industry sectors. There are also thermoelectric plants that are fuelled by burning biomass waste such as firewood/bagasse and even urban waste. It should also be noted that the main fuels used worldwide are coal, oil derivatives and, increasingly, natural gas.

According to the International Energy Agency (2003), the main environmental impacts caused by electricity generation in the atmosphere come mainly from the burning of fossil fuels (oil, coal and natural gas), the so-called greenhouse gases. The most problematic of these gases are carbon dioxide (CO_2), methane (CH_4) and nitrous oxide (N_2O).

Much of the climate change that has been taking place on the planet in recent decades, such as the rise in average temperatures, has been caused by the increase in the concentration of these gases in the atmosphere. One of the main consequences of global warming on the planet is the melting of glaciers, which raises sea levels, flooding coastal areas and islands, affecting large numbers of people and wild animals and transforming the biodiversity of these regions around the world (INTERNATIONAL ENERGY AGENCY AGENCY, 2003).

Renewable energy offers the world a number of advantages, such as low prices, which have fallen considerably in recent years in the case of wind and solar power, as well as minimal environmental problems, especially in developing countries that have high electricity costs. In particular, the use of these renewable resources provides great benefits, especially as they have little environmental impact (HINRICHIS; KLEINBACK; REIS, 2011 apud ALEXANDRINI et al. 2014).

Renewable energies are derived from natural circuits for converting solar radiation, which are the main sources of almost all the energy available on earth and are therefore practically inexhaustible and do not change the planet's thermal balance (PACHECO, 2006).

With a view to global trends in clean energy production, which are mainly related to environmental issues, Brazil has become interested in researching other energy sources to diversify its energy matrix, to the extent that the country can continue to grow in the face of the necessary energy demand, and in this case we can highlight the country's high potential for solar energy generation, as it has the largest reserves of the raw materials needed (ALEXANDRINI et al. 2014).

According to Goldemberg and Lucon (2007), the large emissions of local pollutants and greenhouse gases are generated by current patterns of energy consumption and production, which in turn are based on fossil fuels. This jeopardises the planet's long-term supply. It is necessary to change this scenario by stimulating the use of renewable energy, and in this sense Brazil has a positive position in relation to other countries.

Against this backdrop, Brazil's energy matrix has taken on a new context compared to other countries, becoming increasingly less dependent on non-renewable energy sources such as oil and natural gas (LIMA, 2012).

Table 1 Primary energy in Brazil and worldwide.

Primary Energy				Brazil	World
Total, billion **Tep**				0,193	10,7
Share of Sources (%)	Non-renewable	Fossils	Oil	43,6	35,3
			Natural gas	6,6	20,9
			Coal	6,8	24,1
		Nuclear		1,8	6,4
			Subtotal	58,7	86,6
	Renewables	Traditional	Biomass Traditional	19,0	9,4
		Conventional	Hydraulics	15,3	2,1

	Modern, "new"	Modern Biomass	6,9	1,2
		Others: Solar, wind, etc.	<0,1	1,7
		Subtotal	41,3	14,4

Source: Goldemberg and Lucon (2007), apud International Energy Agency (IEA).

Table 1 shows that Brazil stands out from other countries for one simple positive aspect: the Brazilian energy matrix is already around 40 per cent renewable, compared to the world average of around 15 per cent. In this context, it can be said that the country therefore has a unique opportunity to become one of the world leaders in the energy sector. Boosted by its great hydroelectric potential and with less use of non-renewable energy sources, the country stands out from the rest by reducing environmental impacts in this sector.

As Brazil is a country with many rivers, it has become a strong alternative for generating energy through hydroelectric power stations. According to Reis et al. (2005 p.241), "regardless of the new direction of the electricity sector, this type of generation will continue to have a majority share in view of the enormous potential still to be explored".

However, Goldemberg and Lucon (2007) point out that although Brazil has a strong hydroelectric base in its electricity matrix, it has a very early start compared to the world average in using other modern sources of renewable energy, even with the efforts of the federal government through programmes and incentives in this sector.

According to various authors, due to the environmental problems caused by fossil fuels, many countries are looking for alternative renewable sources of energy generation. In this scenario, Brazil is no different and is seeking to expand its energy matrix by investing in the use of renewable energies as a way of having the least possible impact on the environment.

According to Pacheco (2006), there are a number of renewable energy sources that are essential to the country's energy matrix and which should be used in such a way as to have the least impact on the environment:

a) Solar energy: energy that comes from the sun. It can be used directly to heat the environment, water and to produce electricity, potentially reducing conventional energy consumption by 70 per cent. In addition, it can be converted directly into electrical energy through material effects such as thermoelectric and photovoltaic. Brazil has an excellent solar radiation index, especially in the Northeast, specifically in the semi-arid region, where the best indices are found, with typical values of 1752 to 2190 KWH/m^2 , per year of incident radiation.
b) Wind Energy: The kinetic energy of air masses (winds) caused by the uneven heating of the Earth's surface. This energy has become one of the great alternatives in the energy

matrix of several countries, and in Brazil, this source has been a strong alternative for generating energy in the Northeast region. It is clean, renewable and available everywhere. The largest installations are in the states of Ceará, Pernambuco, Minas Gerais and Paraná.

c) Biomass: Used to make biofuels such as biodiesel. It is energy produced by plants in the form of carbohydrates through photosynthesis. Plants, animals and their derivatives are biomass. Forms of biomass used as fuels include wood, agricultural products and waste, forestry waste, animal excrement, charcoal, alcohol, animal oil, vegetable oil, lean gas and biogas. Biodiesel is a renewable, biodegradable and environmentally friendly fuel. From an economic perspective, the viability of biodiesel is related to the replacement of diesel fleets, enabling a reduction in the emission of polluting gases and sulphur, which will avoid public health costs and the formation of the greenhouse effect, bringing environmental advantages.

Bearing in mind these issues of the relationship between energy and the environment, it can be said that: "Energy efficiency is undoubtedly the most effective way of simultaneously reducing costs and local and global environmental impacts" (GOLDEMBERG and LUCON, 2007 p.18).

2. 3International Energy Efficiency Panorama

According to the National Energy Efficiency Plan (PNEE, 2011 p. 01), energy efficiency is defined as follows:

> Energy Efficiency (EE) refers to actions of various kinds that culminate in the reduction of energy needed to meet society's demands for energy services in the form of light, heat/cold, power, transport and use in processes. In short, it aims to meet the needs of the economy with less use of primary energy and therefore less impact on nature.

Energy efficiency has been developing more and more in international industries, and in this panorama, a summary of the document "International Experiences in Energy Efficiency for Industries" can be presented as a highlight. This document was produced through a partnership between the National Confederation of Industry (CNI) and ELETROBRAS, brokered by PROCEL - the National Electricity Conservation Programme and its PROCEL INDÚSTRIA sub-programme (2012).

In this study, which surveyed several countries, industrial energy efficiency programmes were identified that had obtained the best results and could possibly serve as a benchmark for Brazil, some of which are highlighted below:

2.3. 1United States

The United States has developed several programmes aimed at energy efficiency, at federal, state and electricity and natural gas company level. These actions are important at the national level, such as the Industrial Technology Programmes, the general objective of which is to reduce the energy intensity of industry by 30% by 2020 and to implement more than 10 efficient technologies by 2010 through public-private partnerships in research and development (R&D) projects.

These programmes aim to meet specific energy efficiency objectives such as:

- Forming partnerships between private companies in R&D projects aimed at reducing energy costs and increasing competitiveness in some energy-intensive industry segments (aluminium, pulp and paper, metal smelting, chemicals, mining, oil refining and steel);
- Saving energy in the most intensive industrial processes with more efficient equipment;
- Investing in industry-oriented strategies, through energy diagnostics and the dissemination of best practice information in terms of motors, pumps, fans, direct heating, process steam and compressed air.

2.2.1 Canada

Canada has three main segments involved in this management. The Office of Energy Efficiency (EEE), the Varennes Technology Centre (CTEV) and the Energy Efficiency Agency, both linked to the Ministry of Natural Resources.

The EEE also runs a voluntary programme between associations and companies, applied in various industrial sectors, and is also responsible for implementing energy efficiency programmes, working with the EcoEnergia programme for industry, encouraging companies to adopt more efficient processes through:

- Renovations to industrial facilities, aimed at reducing utility costs such as energy and water, through the incentive of 10 Canadian dollars per GJ of energy saved, for companies with up to 500 employees that do not operate in sectors with GHG (Greenhouse Gas) reduction targets;
- Integration of processes that increase the energy efficiency of industrial facilities and reduce GHG emissions. For this, there are incentives of 50,000 Canadian dollars. The amount is calculated on 50 per cent of net invoiced costs, exempt from refunds, subsidies and other incentives. In these cases, the industrial plant must have independent energy and water networks and the water treatment or energy supply systems must be at their capacity limits.

CTEV seeks to identify potential in industries for reducing specific consumption and to develop tools to help achieve these goals, especially in the pulp and paper, oil refining, chemical, steel and food industries.

2.2.2 France

France has two bodies focused on energy efficiency, the Environment and Energy Management Agency (AMAGE) and the General Directorate for Energy and Raw Materials (DGEMP). The country also offers technical and financial support to companies seeking energy efficiency. Under various international agreements such as the Kyoto Protocol, France aims to reduce emissions by 15 per cent between 2008 and 2012 compared to 2007 levels.

AMAGE supports research and the development of tools and diagnostics aimed at energy efficiency, as well as disseminating information, offering tax incentives and professional training in various sectors:

- Acquisition of information to gain a better understanding of industrial energy consumption, with the aim of identifying savings opportunities in this sector;
- Investment in new energy efficiency technologies;
- Information on separation, drying, compressed air, ventilation, generation and refrigeration processes;
- Implement energy-saving solutions by offering limited funding of up to 30,000 euros, which can cover up to 50 per cent of the costs;
- Incentives for projects to replace equipment, improve facilities and use renewable energies with the help of funding of up to 750,000 euros, which can cover up to 70 per cent of the total cost.

The DGEMP defines national energy policies to guarantee the country's supply from mineral sources. Its main tasks are: Analysing the prospects for the evolution of the country's energy matrix; Coordinating research into renewable energy sources, fossil fuels and nuclear energy; Drawing up regulations to increase the energy efficiency of equipment. It also issues an Energy Saving Certificate which obliges energy suppliers and distributors to encourage companies to save energy, under penalty of a fine in the event of non-compliance.

2.2.3 England

A private company called the Carbon Fund has been set up in England to combat climate change. This association works together with the public/private sector to develop technologies to reduce carbon emissions. Energy efficiency actions are primarily focused on training, information dissemination and energy diagnostics.

Tools offered by the Carbon Fund:

- Development of new technologies that reduce carbon emissions, by financing projects;
- Evaluation of carbon emission strategies, informing the government about actions and

trends in this sector and seeking to change business thinking in this regard;

- Practical advice for the public and private sectors on finding concrete solutions to reduce carbon emissions and electricity consumption;
- Creating new companies aimed at reducing carbon emissions by financing the best ideas and promising business plans.

2.2.4 Japan

Japan has set up an Energy Conservation Centre, which carries out the following actions:

- Keidanren Voluntary Environmental Action Plan: prepares energy diagnoses and disseminates information through practical guides on the efficient use of energy for energy-intensive industries, with a focus on steel, chemicals, paper and electricity;
- Energy Audits in Factories: formulates free energy diagnoses in companies in order to identify improvements in energy issues and advise them on the best use of available resources;
- Financing projects that seek to minimise energy consumption and cogeneration of this resource.

2.2.5 China

China has a series of serious environmental problems because it has the highest intensity of electricity consumption among industrialised countries. The industrial sector is responsible for around two thirds of the country's primary energy consumption.

The China Energy Group (GEC), created by the Lawrence Berkeley National Laboratory in the United States, exists in China. This group is made up of researchers and suppliers from China and various countries, with the aim of gaining a better understanding of Chinese energy consumption and improving the capacity to create improvement programmes and tools for this sector.

The GEC also develops programmes for energy efficiency in the driving forces, and encourages the formalisation of voluntary agreements aimed at energy efficiency in industry.

China has a National Development and Reform Commission, which, in partnership with the United Nations Development Programme, has been running the End-Use Energy Efficiency Programme since 2005, aimed at more efficient management of energy resources and more sustainable development through the use of renewable energy technologies.

China also has small mandatory energy efficiency standards for various pieces of equipment used in industry, such as electric motors, pumps, air compressors and fans.

More and more Chinese companies are adopting optimised energy management standards in their industrial processes, combined with ISO 9000 and 14000.

2.4 Brazil's Energy Scenario

As pointed out in section 2.3 of Chapter 2, there is no worldwide standard for actions and measures aimed at energy efficiency in industry. The bodies and associations involved in managing this input are developed through public-private partnerships, which seek to implement energy efficiency programmes aimed mainly at energy-intensive industries.

In this panorama, Brazil has significant experience with programmes aimed at energy conservation in industry, but the use of these mechanisms has many discontinuities, necessitating the implementation of various changes between government and industry in order to increase the productivity of public resources used in industrial energy efficiency programmes and also attract private capital for new programmes (PROCEL, 2012).

In Brazil, the main regulatory standard in this sector was the enactment of Law 10.295/2001, known as the Energy Efficiency Law, created by the government due to concerns about electricity consumption, which ultimately sets out the country's national energy conservation and use policy (PNEE, 2011).

It is also worth pointing out that the Brazilian government has never had a consistent energy policy aimed at energy efficiency, with targets and energy conservation agreed between its main players. A policy drawn up with a chapter on energy efficiency in industry would be the federal government's first step, in the field of the National Energy Policy Council (PROCEL, 2012).

In Brazil, there is still no body designated to plan and manage industrial energy efficiency actions. The MME, through the Secretariat for Energy Planning and Development, employs Procel and the National Oil and Natural Gas Conservation Programme, managed simultaneously by ELETROBRAS and Petrobras. These two programmes have recently shown modest results in terms of energy efficiency (PROCEL, 2012).

In this context, according to PNEE (2011), it can be noted that in Brazil various actions have been explored for the rise of energy efficiency in industry. Specifically highlighted are the PROCEL industry programme (National Electricity Conservation Programme); PROESCO (Support for Energy Efficiency Projects) with a line of financing from the National Bank for Economic and Social Development - BNDES; Energy Efficiency Programmes - PEE, run by the National Electricity Agency -ANEEL, and the National Programme for Rationalising the Use of Petroleum Derivatives and Natural Gas - CNPET, run by Petrobras.

Among the actions to standardise energy efficiency is the Special Study Commission on

Energy Management, translated from the ISO 50001 standard by the Brazilian Association of Technical Standards (ABNT/CCE, 2009), which establishes the systems and requirements needed to improve energy performance, including energy efficiency and energy intensity within organisations, such as the implementation of an energy policy with objectives, targets and action plans, taking into account legal requirements and information regarding the efficient use of energy. This standard applies to all sizes of organisation and helps to drive down costs, greenhouse gas emissions and other environmental impacts through systematic energy management.

Most countries have mandatory legislation and regulations for energy efficiency planning in industries. In Brazil, the Energy Efficiency Law allows minimum and maximum levels of efficiency or consumption to be set for equipment, vehicles and buildings. Electricity supply companies carry out EEPs that are fine-tuned by ANEEL, which has been prioritising the use of projects included in EEPs, but does not set energy conservation targets (PROCEL, 2012).

"On 24 June 2000, Law No. 9.991 was enacted, which regulates the obligation for Brazilian electricity distribution companies to invest in end-use energy efficiency programmes" (PNEE, 2011 p.17).

Several countries have quantitative targets for subsequent gains in energy efficiency, adopting minimum and mandatory standards in this sector with the support of voluntary agreements between government and industry. In Brazil, the Strategic Energy Efficiency Plan calls for 0.5% of the net revenue of electricity distribution companies to be invested in EEPs. The PNE 2030 set the conservation target at 106 TWh/year, but the EEP, which would have to implement the actions and programmes needed to meet this target, has not yet been properly implemented (PROCEL, 2012).

Another aspect in Brazil in relation to EE is the lack of culture in organisations, due to investments in energy performance improvements, demonstrating a short-term vision in this sector. In addition, some companies have a small workforce with difficulty in seeing energy efficiency as an organisational strategy (PNEE, 2011).

There is also a lack of professionals and technicians qualified to work in EE in the industrial sector, both in the equipment used and in industrial processes. This deficiency is pointed out in the lack of information and specific knowledge about EE on the part of the engineers and technicians who work in this sector (PNEE, 2011).

The country's energy efficiency programmes will only be successful if effective measurement and verification mechanisms are implemented in all programmes, with a plausible degree of decentralisation in relation to the states of the federation. Another way would be to implement optimised energy management standards in industry in some energy-intensive segments, compatible with ISO 9000 and 14000, and the formation of public-private partnerships for R&D projects with efficient industrial equipment and processes is an essential strategy for achieving energy efficiency gains that is little used in Brazil (PROCEL, 2012).

According to Sola and Kovaleski (2004), the competitiveness of companies in the electricity sector has led to a privative approach to their actions: the organisation that sells and supplies electricity does not have the objective of rationing consumer demand.

According to Goldemberg (2000), most of the equipment used in transport, homes and industries was designed at a time when energy was plentiful, cheap and there was little concern for the environment. These aspects show why today there are many opportunities to improve consumption in the electricity sector.

Compared to other countries, Brazil has a privileged hydrography due to a series of natural aspects, such as climate, geomorphology, topography, among others. It also has an abundance of water, which means that these resources can be better utilised for the production of electricity (FARIA, 2011).

The unit of power in an energy system Watt (W) means the energy of 1 joule per second, i.e. a Whatt-hour is the amount of energy needed to power a load with a power of 1Whatt for 1 hour. The power rating on each appliance indicates its power for a period of one hour. So an appliance with a power of 5400 W (Watts) will have the equivalent consumption of this power in 1 hour (ATLAS, 2008). Other multiples of Watt-hours can also be mentioned: (a) Megawatt-hour (MWh) equals 1,000,000 Wh, (b) Gigawatt-hour (GWh) equals 109 Wh, (c) Terawatt-hour (TWh) equals 1012 Wh.

As a country rich in water resources, hydroelectric power has always been dominant. The installed capacity of this type of energy in Brazil is around 70,000 megawatts (MW, million watts). According to Goldemberg and Lucon (2007), there are 433 plants in operation and only 23 of these have a capacity greater than 1,000 MW, which represents 70% of the total installed capacity. The Amazon region is where most of the unused potential is concentrated, around 190,000 MW. It should also be noted that it costs approximately US$1,000 to produce 1kW of power in a hydroelectric plant.

An energy matrix is a description of all the energy produced and consumed in the country, broken down by production source and consumption sectors (VICHI and MANSOR, 2009). In Brazil, the detailed availability of the matrix is through the National Energy Balance (BEN), which is drawn up annually by the Energy Research Company (EPE) and published by the Ministry of Mines and Energy (MME).

According to the National Energy Balance (2013), electricity generation in Brazil totalled 552.5 THW (Terawatt-hours) in 2012, an increase of 3.9% compared to 2011, with the main contribution still coming from public utilities with 85.9% of the total and 16.7% of generation coming from non-renewable energy, with self-producers accounting for 14.1%. Final consumption of the energy supplied in the country totalled 498.4 TWH, an increase of 3.8% on the previous year.

According to the National Energy Balance (2014), hydroelectric power generation is predominant with 70.6% of the total generated, followed by natural gas with 11.3% and biomass 7.6%, among others, totalling 609.9 TWH of energy generation from all renewable and non-renewable sources in 2013 Figure 1. It is estimated that by 2030 this consumption will be between 950 and 1250 TWH (ANEEL, 2007).

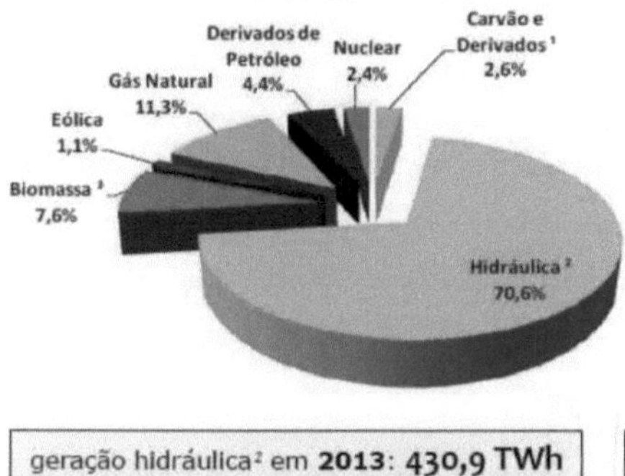

Figure 1 Brazil's electricity matrix.
Source: MME (2014).

As shown in Figure 1, Brazil is becoming more efficient at producing clean energy, which is a positive point in economic and environmental terms, since hydroelectric power is less polluting than other sources such as nuclear power. Another aspect can be seen in the consumption estimates for the coming years, which are expected to rise sharply and will require major future government investment in this sector.

Table 2 Installed Power Generation Capacity.

REGION	HYDRO%	TERM%	WIND%	NUCLEAR%	TOTAL%
NORTH	15,1	12,7	0,0	0,0	14,0
NORTHEAST	13,4	22,2	64,0	0,0	16,4
SOUTHEAST	29,4	40,8	1,5	100,0	33,3
SOUTH	28,4	14,8	34,5	0,0	24,3
WEST CENTRE	13,6	9,5	0,0	0,0	12,1
TOTAL	100,0	100,0	100,0	100,0	100,0

Source: BEN (2013).

Table 2 shows that the south-east is responsible for the largest energy generation capacity in the country, with 33.3 per cent of the total, followed by the south, which accounts for 24.3 per cent. Another aspect is nuclear power generation, which is only produced in the south-east, which is a positive point because it is concentrated in only one part of the country, since this type of generation can cause major pollution and environmental impact if not used and controlled correctly. In terms of wind power generation, the southern region leads this segment with 34.5 per cent of the country's production, which is another positive aspect, as this type of energy generation is considered clean and contributes to preserving the environment.

2. 5Industries' Energy Scenario

Despite the increase in electricity consumption in the residential and commercial sectors, with growth of 1.3 per cent more than Brazil's GDP, the industrial sector is still the largest consumer of electricity with 34.4 per cent of the country's domestic energy supply, followed by the residential sector with 20.5 per cent and the commercial sector with 13.8 per cent, among the other consumer sources (BEN, 2014).

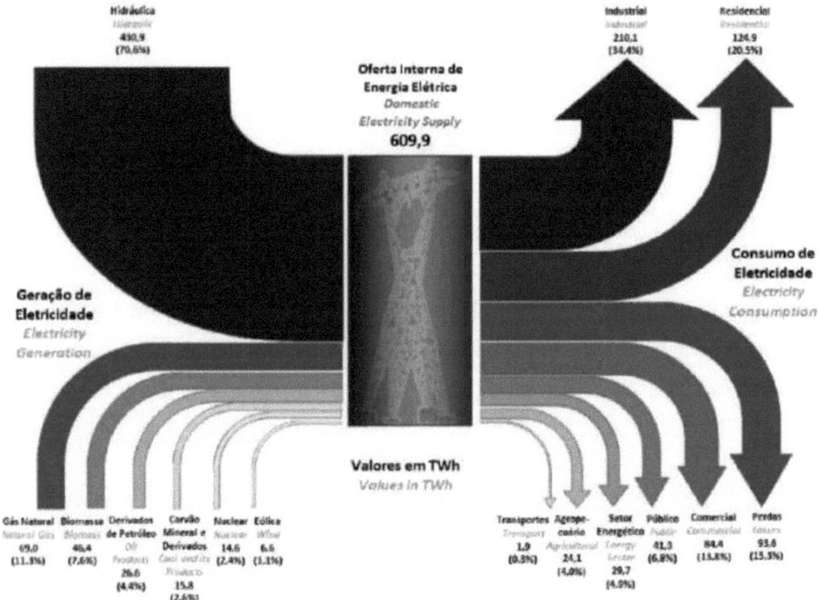

Figure 2 Brazil's electricity matrix.
Source: MME (2014).

It can be seen that despite the participation of other renewable and non-renewable sources in electricity generation such as wind, nuclear, biomass, natural gas, coal and oil derivatives, hydropower is still predominant in the domestic electricity supply Figure 2. The industrial and residential sectors still lead the consumption indices, and it can be seen that the number of energy losses is significant in relation to the other sectors, even surpassing the commercial sector, as it represents 15.3 per cent of the energy offered by the matrix.

Sola and Kovaleski (2004) point out that the National Energy Balance shows that a large proportion of energy losses are related to obsolete equipment and processes that are still used in industries today. From an economic, operational or environmental point of view, these losses cause major damage to the entire production chain. Industries that have an Integrated Management System are more efficient and aware of losses, and also have a more consistent energy policy, so that within of these companies, a culture of energy efficiency is created through a permanent policy.

"In industry, the non-ferrous sector, where the aluminium industry stands out, accounts for almost 20% of electricity consumption. The chemical sector accounts for 12.4% and the food and drink sector for 12.6% of this consumption" (PNEE, 2011 p.29).

Industrial energy consumption has been growing, thus requiring investment in generation, distribution and transmission to meet the sector's increased electricity supply. The implementation of programmes, projects and activities aimed at the efficient use of this energy by the various industrial segments should be a constant stimulus, given the major challenges facing the electricity sector in meeting demand across the country (CNI, 2014).

If this profile of electricity consumption by industries is fuelled, environmental damage and impacts tend to be constant, partly due to the need for major projects in the electricity sector to meet demand and the increasing extraction of natural resources (LESTINGUI et al., 2009).

As such, there is a need for investment and the implementation of public policies with concrete goals and objectives aimed at reducing energy consumption in this sector, through measures that encourage the modernisation of production plants and the emergence of innovations that can reduce energy consumption in their production processes (LESTINGUI et al., 2009).

Companies interested in reducing electricity consumption in their production processes and improving the energy efficiency of their installations, in a way that does not compromise safety, product quality or production capacity, can obtain the following advantages: Improved utilisation of electrical installations and equipment, with a consequent improvement in product quality; reduced energy consumption and a consequent increase in productivity, without affecting safety; reduced electricity costs (COPEL, 2005).

According to Alves et al. (2007), reducing electricity consumption in industry can result in both a reduction in costs, due to the lower energy demand required, and a gain in productivity, since when the organisation maintains the contracted energy demand, it takes advantage of this input, which would otherwise be spent on electricity, to invest in new equipment as a way of expanding its production.

According to PROCEL (2005), an example of energy efficiency actions is a successful case of a textile industry located in Pacajús - CE, which based on an energy diagnosis, established improvement actions, highlighted as:

- Adaptations to modernise the lighting systems in the industrial warehouses, in which HO

110W lamps and 2 X 110W double electromagnetic ballasts predominate;

- Use of translucent roof tiles in the industrial warehouses and also in various other sectors. This investment has led to greater use of natural lighting, with a consequent reduction in artificial lighting during the day;
- Optimisation of the air and water flow rates in the central air units and refrigeration machines in industrial warehouses, by adjusting the automation and temperature and humidity control systems, with the aim of reducing energy consumption in the chilled water generation systems;
- Modulation of peak-time loads in the wiring preparation sector, concentrating services during off-peak hours, which allowed equipment to be switched off at this time;
- Retrofitting the fan blades in the spinning and weaving air centrals (air conditioning) by replacing the 8-bladed blades with similar 5-bladed blades. Significantly reducing the energy load required by halving the number of fans.

These actions totalled R$196,100.00. With this modernisation, the industry saw a 2.3% increase in production in the same year, and 40% in subsequent years with new implementations in other industrial parks linked to the company. It also saw a 13% reduction in electricity costs, resulting in 7,279 MWh/year in electricity savings.

According to CNI (2014), in order to conserve electricity within industries, an awareness campaign must first be carried out, motivating all employees by distributing leaflets, posters, manuals and news items in internal newspapers.

According to Ferreira et al. (2009), the majority of diagnoses carried out in Brazilian industries recommend replacing several electric motors, often accompanied by adapting the lighting system. Usually, the adaptation of lighting systems is linked more to reducing electricity costs than to reducing energy consumption.

In recent years, PROCEL Industry has been working with state industry federations and universities to develop actions on motor systems. These involve monitoring machinery and equipment, measuring power, recording motor data and proposing that the machine be replaced with a high-efficiency one, taking into account that the load driven by a motor is constant (FERREIRA et al. 2009).

The main focus of electricity consumption in the industrial sector is related to the use of motive power, which includes electrical consumption in equipment such as pumps, fans, compressors, in various applications such as fluid and gas processing, refrigeration, among others (PNEE, 2011).

According to Copel (2005), energy conservation currently needs to be established as one of the main objectives of any well-managed industry. In this alignment, numerous alternatives and measures are required, such as the adoption of operational and administrative measures, as well as the creation of parameters and control and monitoring, among others.

As an administrative measure for energy conservation, equipment and machines should be chosen that take energy efficiency into account, emphasising the choice of models that preferably have lower losses or lower specific consumption to carry out the same activity (COPEL, 2005).

Measures such as the technical specifications of manufactured products, with the aim of verifying the possibility of reducing the use of energy in their manufacture, are also actions aimed at conserving energy. Improving processes and controls to ensure better product quality is also an indirect way of conserving energy (COPEL, 2005).

According to the ABNT/CCE standard (2009), when purchasing energy services, products and equipment that use a significant amount of electricity, the organisation must emphasise to its suppliers that the purchase is based on energy efficiency. In this way, the company can define specifications when purchasing energy based on the following items: energy quality; availability and capacity; variation over the specified time; billing parameters; costs; environmental impact and renewability.

Activities such as the planning, scheduling and execution of maintenance are very important measures in industry, which can bring significant results in energy conservation if they are carried out properly. Furthermore, the malfunctioning of equipment, machines and installations is a consequence of inefficient maintenance and usually causes excessive energy consumption (COPEL, 2005).

Monitoring energy bills on a monthly or weekly basis is basic for controlling expenditure on this input. However, when organisations have large facilities, this analysis can be difficult, making it advisable to monitor them at various points by installing meters in various locations, such as sections, warehouses, circuits or even machines (COPEL, 2005).

This method not only makes it possible to control electricity consumption and expenditure, but also to identify the way in which energy is consumed, as well as identifying the points where action should be taken to reduce electricity consumption (COPEL, 2005).

In this way, organisations that have an internal team or committee that is responsible for monitoring, identifying and implementing actions and solutions for the reduction and efficient consumption of electricity also stand out as being able to control energy expenditure. One example is the CICEs (Internal Energy Conservation Commission), which are formalised commissions in the Federal and State Public Administration that aim to implement and monitor effective measures for the rational use of electricity, as well as controlling and disseminating the most relevant information. Their concept applies to all types of facilities, whether they are in the federal, state or municipal sectors. (CICE, 2015).

A well-prepared project and plan for the operation and maintenance of electrical installations can lead to significant energy savings, as well as maintaining good operating conditions and equipment safety, guaranteeing the continuity of production (COPEL, 2005).

Still in the present context of actions and measures for energy efficiency in industries, the ABNT/CCE (2009) standard, which establishes the requirements and systems for energy management, defines the goals and tools for continuous improvement necessary for organisations to be able to set energy efficiency objectives and targets and thus obtain certification.

According to ABNT/CCE (2009), organisations must develop, maintain and record an energy profile. This is done by analysing energy use based on measurements of other data, identifying current and potential energy sources, evaluating past and present energy use and estimating future energy use. In addition, it is necessary to identify the facilities, equipment, systems, processes and employees that significantly affect energy use, as well as other variables that affect energy use, in order to determine and prioritise opportunities for improving energy performance, including the use of renewable or alternative energy sources.

"Energy performance is a broad term that includes energy efficiency, energy intensity, energy consumption, and so on. Energy efficiency is a basic concept of energy performance" (ABNT/CCE, 2009 p.15).

With regard to objectives, targets and action plans, the standard states that organisations must set, implement and maintain documented energy objectives and targets, which must be measurable, have a deadline for implementation and be compatible with the company's energy policy, demonstrating commitment to progress in the energy sector and complying with the company's legal obligations (ABNT/CCE, 2009).

This documentation must contain: the energy policy; the energy objectives, targets and action plans; plans for achieving the objectives and targets; documents and records required by the standard and the documents and records determined by the organisation that are necessary for planning and controlling the processes and equipment directly involved in the use of identified energy (ABNT/CCE, 2009).

The energy policy is the organisation's management's official statement of obligations regarding energy management. This policy contains the processes for continuous improvement, the availability of information and the necessary resources established by the organisation in order to meet the objectives, targets and legal requirements that apply to energy performance (ABNT/CCE, 2009).

"The policy may include additional commitments such as alternative energy sources, reduced environmental impacts related to energy and renewable energies" (ABNT/CCE, 2009 p. 18).

CHAPTER 3

METHODOLOGY

This chapter will present the approach and methodology used in the research, the procedures adopted to obtain and analyse the data, its delimitations, the scenario studied, the objectives and also the research instrument used.

3.1 Approach Method

According to Gil (2010), the methodology is important for evaluating the results of the research and showing how the data was obtained, as well as the stages in which the study was carried out.

In order to achieve the objective proposed in this research project, the deductive method was used, which starts from a broader knowledge to certain particular cases, which in this case aims to have a general conclusion about the phenomenon investigated (DINIZ and SILVA, 2008).

3.2 Research Classification

3.2.1 In terms of nature

Regarding the classification of the research and its nature, it is applied because it is aimed at acquiring knowledge that can be applied to a specific occasion (GIL, 2010). In this study, the aim is to investigate how industries in Ponta Grossa act in terms of energy management.

It also focused on understanding the processes used in organisations in relation to electricity consumption, as well as identifying the energy sources used, their techniques and improvement actions aimed at this aspect.

3.2.2 As for the problem

With regard to the problem, a quantitative approach was used where the variables studied and the information acquired can be quantified, allowing the use of correlations and other statistical methods. The quantitative characteristic also refers to the technical procedure employed, which requires the use of statistical instruments as a basis for analysing the problem. The data obtained through a survey can be grouped in spreadsheets or tables, thus improving its analysis (GIL, 2010).

3.2.3 As for the objective

With regard to the objectives, this is a descriptive study that involves collecting data through an analysis tool in order to obtain a broader knowledge of the subject, which aims to detail the phenomenon observed step by step, collecting the information necessary for examining and concluding the energy management scenario of the industries in Ponta Grossa.

3.2.4 Regarding the technical procedure

With regard to the technical procedure of the research, a survey was adopted using a questionnaire to question people from the chosen companies, in order to collect data and information pertinent to analysing and concluding the scenario under study.

According to Gil (2002), the survey seeks direct information from a group of interest, in order to achieve the description of the specific objectives of the study, in the clearest possible terms. This is done through the use of questioning techniques such as questionnaires, forms or interviews, in which an individual is directly involved so that concrete data and information can be obtained, through a deep and exhaustive study of one or a few objects, in such a way as to allow a broad and detailed knowledge of the subject studied.

A bibliographical approach was also used, based on material published in books, articles, newspapers and theses on the subject, as well as documents issued by the federal government on the National Energy Balance, the National Energy Efficiency Programme and comparisons of estimates by the Ministry of Mines and Energy.

3.3 Limitations of the work

Regardless of the instrument used in a survey, questioning techniques make it possible to acquire data from the point of view of the interviewees. In this way, the survey will always have limitations when it comes to studying broader social relations, especially when it involves variables of an institutional nature (GIL, 2002).

As far as the limitations of the survey method are concerned, one of them relates to the fact that the Ponta Grossa region has many companies of different sizes and branches of activity. The research focused only on the city's main companies, which play a significant role in the region's economy.

On the other hand, this is an important issue at company level, since companies that have efficient energy management can achieve significant results in terms of cost reduction and their corporate image, since it is also an environmental and social issue.

As far as the data collection instrument is concerned, the main limitation is the guarantee that the interviewees from the organisations will answer the questionnaire. Also, the availability of time given by them to interrupt their activities for the interview and explain the information, limitations which imply a reduction in the sample.

3.4 Population and Sample

As a research technique, the procedure will be sampling, which aims to identify in a population a portion of the group to be studied (MASCARENHAS, 2012).

In the case of a survey, the population that makes up this research is made up of the main industries in the region, which according to the Commercial, Industrial and Business Association of Ponta Grossa (ACIPG, 2015) are as follows: CCR Rodonorte, Masisa do Brasil, BRF Brasil Foods, Bunge Alimentos, Coinbra, Heineken, Hubner, Insol do Brasil, Continental/Contitech, Cargil, Stora Enso, Tetra Pak, Winner Chemical, Lojas MM, Sudati, Braspine Madeiras, Branslumber Indústria de Molduras, Geroma do Brasil, Makita, Kemira Chemicals, W3, Kurashiki, Louis Dreyfus, Macrofértil Fertilizantes, Beaulieu do Brasil, Águia Sistemas.

These 26 organisations stand out because most of them are multinationals with a major stake in the economic, social and cultural development of the region and the state of Paraná.

Because a survey covers a universe of elements with large numbers, it is difficult to measure their totality. For this reason, it is best to work with a sample, i.e. a small part of the population. When the sample is selected with a certain rigour, the results obtained tend to be very close to the whole population (GIL, 2002).

According to Gil (2002), the sample of a population in a study is made by selecting functions that have characteristics that make it possible to verify a certain display in the occurrence of a particular phenomenon. In this study, a simple random probability sample was taken, which according to Maldim (2004, p. 113), "is the basic method of random sampling, due to its ease of selecting samples, analysing data and reducing sampling errors".

According to Maldim (2004), the phases of the simple random sample method are listing the population, determining the sample size and using random numbers. Based on this, the method used was to list the population (26 companies), the sample size is approximately 40% of the population, each numbered and chosen at random. This resulted in 10 companies in the region, classified as medium or large, in which the case study was applied to the selected organisations.

The companies and interviewees in each sample are not identified in this study in order to preserve the corporate image of the organisations, also due to their operational strategies and for better cooperation on their part in providing data and information for this analysis.

3.5 Research Instrument

"The formulation of a questionnaire basically consists of translating the specific objectives into well-written items" (GIL, 2002).

In this study, the data collection instrument is a questionnaire with 17 questions, 15 of which are closed and 2 open, as shown in Appendix A. The questionnaire was drawn up on the basis of a literature review, with the aim of understanding the attitude of industries towards energy management.

Using the theoretical framework in Chapter 2 as a basis, the questionnaire was formulated. It was divided into three blocks, described below:

- Block 1 - This seeks to identify whether organisations have adopted or have a policy aimed at energy management, by analysing internal declarations, certifications aimed at this sector and whether there are written procedures or manuals that meet the requirements on energy use.
- Block 2 - Aims to identify the sources of energy used in the industry and also the equipment that most generates the consumption of this good, through research into the types of energy used, their percentage of consumption and in which systems the highest rate of electricity consumption is registered.
- Block 3 - This seeks to identify additional information for the formalisation of the survey, by measuring the energy efficiency actions used by the companies. To this end, it seeks to investigate the organisations' objectives and goals in this sector, the existence of public-private partnerships, the evaluation methods for purchasing energy and equipment, and the outlook and analysis of future energy consumption.

A pilot test is necessary to evaluate the instrument before it is applied in order to: develop the application procedures; test the vocabulary used in the questions; ensure that the questions set out in the questionnaire allow the variables being studied to be measured (GIL, 2002).

In order to find out whether the questions in the questionnaire are coherent with the topic of energy management in companies, the instrument was analysed to determine whether the questions are coherent with the topic of energy management in companies.
validated by three managers from industries in the region who have specific knowledge of the subject.

Secondly, the questionnaire (Appendix A) was applied as a pilot test to academics on the logistics course at this institution in order to assess the vocabulary of the questions and whether or not it was easy to answer the form.

The questions were inserted into a personalised form using the Google Drive tool, which can be accessed via Gmail in the Google Forms bar and saved on the web.

It was thus possible to send the questionnaire through the system itself, with the answers stored in the email inbox. This tool made it possible to analyse the data in a more efficient way, as the programme itself collects the information and produces reports, spreadsheets and graphs of the responses obtained, allowing for a broader and more detailed diagnosis.

CHAPTER 4

RESEARCH 4.1 Characterisation of the Region Researched

The city of Ponta Grossa has stood out for the large number of companies that migrate to the region, and currently, according to the newspaper Diário dos Campos (2015), with 37,661 active companies, Ponta Grossa ranks 59th out of the 200 Brazilian cities with the most registrations in the National Register of Legal Entities. Updated figures from Empresômetro, a technological tool devised by the Brazilian Institute of Planning and Taxation.

Also according to ACIPG (2015), the city stands out for being the largest metal-mechanical centre in the interior of Paraná, a strong industrial centre for vegetable oils and fats, a large vehicle repair centre, the 15thª largest economy in southern Brazil, 69ª in consumer power in Brazil, 6th best micro and small business development index in Paraná, 6th largest commercial centre in Paraná, 4th largest mineral extraction city in Paraná, 5th in the state ICMS collection ranking and also one of the largest soya crushing centres in Latin America.

4.1 Research Considerations

Once the random sample of companies in the region had been defined, contact was made in August 2015 via email and telephone with the heads of the selected organisations. The work was focussed on the maintenance area, as this sector is totally linked to the subject of the research, and the supervisor or person in charge of the area has knowledge of the subject.

With regard to the research instrument, after the pilot test and validation of the questionnaire (Appendix A), the data was processed in order to remove any answers that were not in line with the desired response.

Emails with the form were sent in October, as the research was only authorised by the ethics committee at the end of September.

The questionnaire was answered without any criticism from the interviewees from the organisations, in terms of the interpretation of the questions and the use of the Google Drive programme used by the interviewer to send the form. In addition, the objective questions were answered in a clear and coherent manner.

The results of the objective questions were tabulated quantitatively using the "Google Forms" programme, which made it possible to extract and organise the data more efficiently.

In general, the reception from the companies was satisfactory, with only two (2) hesitating to take part in the survey and not answering the questionnaire. The first contact established was essential, because at this stage it was possible to inform the organisations and make it clear to them how the survey would work.

Some of the limitations to carrying out the research include the fact that companies are constantly in operation and often do not have the time to analyse and interpret this type of study. Another issue is the difficulty in getting a visit or interview with the people in charge of the organisations, due to the bureaucracy and visitation policies of the companies, as well as the interviewees' fear of providing internal information, even though they stress total confidentiality.

The name of each company and its interviewee has not been disclosed in this study in order to preserve the identity and image of the corporation, since the information obtained has been analysed in a general context.

4.1 Presentation and Interpretation of Data and Results

This topic will present the data and results obtained from the organisations through the analysis of the answers to the 17 questions, which were subdivided into 3 research blocks, divisions based on various authors that enabled a critical and organised diagnosis, achieving the specific objectives of the work (appendix A).

4.1.1 Data and Results from the First Block

Moving on to the analysis of the results, the first block of questions (1 - 4) sought to achieve the first specific objective of the research, which is to identify information on the existence of an energy policy within the organisations.

Question 1 then revealed the segment of companies in the chemical, metallurgical, automobile, textile, wood and beverage industries. This allowed us to analyse various types of industrial activity and not limit ourselves to just one area of activity.

With regard to the second question, which deals with the existence of an official company declaration with commitments on energy management issues, it was found that 50% of the companies have official documentation on this aspect, as shown in Graph 1. This is a partially positive point, as it demonstrates that half of the companies are prepared and concerned about procedures relating to the consumption and use of electricity.

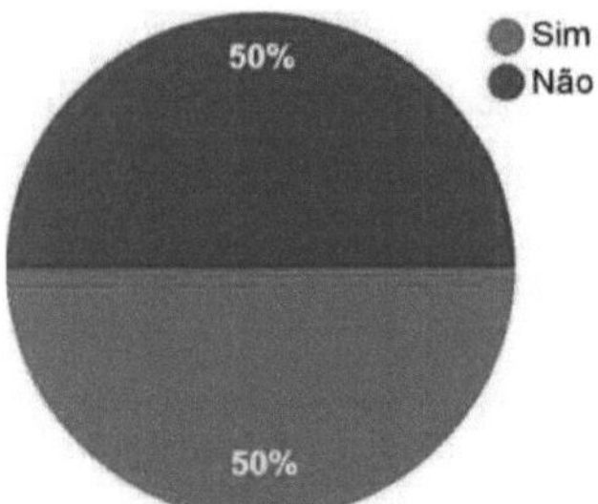

Graph 1 Official Statement on Energy Management Matters.
Source: Prepared by the author.

The third question sought to identify whether the companies have ISO 50001 certification, which establishes the requirements for energy management systems. None of the organisations are certified to a standard related to electricity use systems (Graph 2), thus highlighting a negative point for the region's industries, given that this certification is extremely important as it establishes the basic resources for the efficient use of energy, and also reflects on the companies' good corporate image.

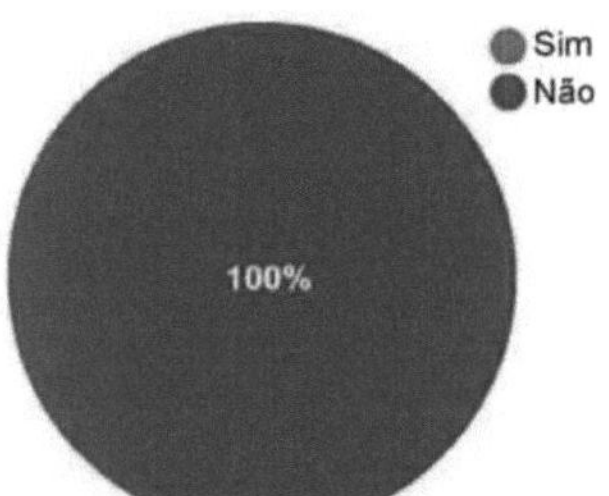

Graph 2 Certification in Energy Management Systems.
Source: Prepared by the author.

On the other hand, question 4, which sought to identify the existence of written procedures relating to the use of electricity, showed that 100 per cent of the companies have or adopt written documents or manuals relating to procedures and the use of electricity. This is a positive observation, as shown in Graph 3.

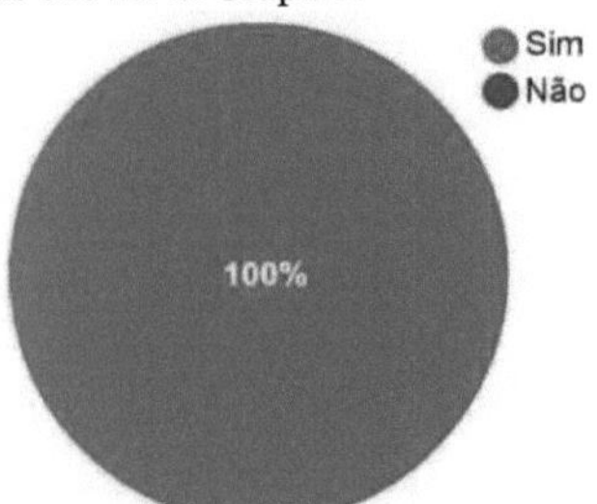

Graph 3 Written procedures on energy use.
Source: Prepared by the author.

4.1.2 Data and Results from the Second Block

With regard to the second block of questions (5 - 9), the aim was to address the second specific objective relating to the consumption and sources of energy used by organisations. This block also included two descriptive questions on the percentage of energy consumed and the equipment that most demands this commodity. These questions are fundamental for identifying the types of mechanisms that most consume energy within industries.

Questions 5 and 6 showed that the biggest source of energy used is electricity, followed by natural gas with 62.5 per cent and oil with 25 per cent. Some companies use other alternative sources such as biomass, coal and biogas, but with much lower percentages, as shown in Graph 4. It can also be seen that electricity accounts for the largest share, with two of the organisations using 100% of this source, as shown in Graph 5.

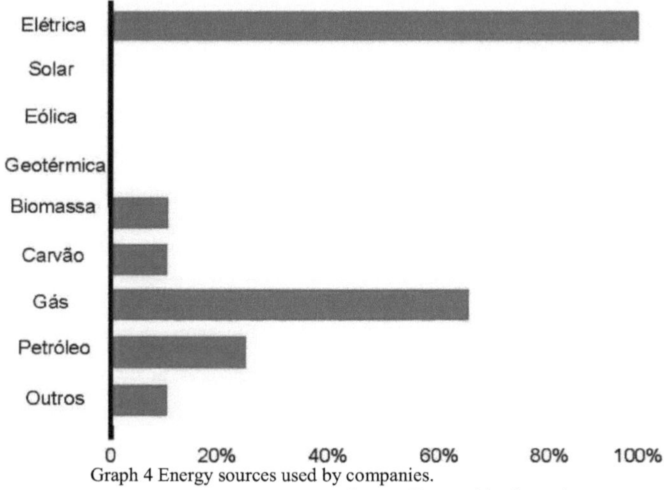

Graph 4 Energy sources used by companies.
Source: Prepared by the author.

It can be seen from these two questions that the companies are very dependent on the electricity source, where as previously mentioned, energy generated by hydroelectric power stations is still predominant in the country. One positive aspect is the use of other sources, such as gas, biomass, oil and coal, diversifying the electricity matrix used. However, none of the industries still use renewable energy sources such as solar and wind power, which are considered modern and sustainable sources.

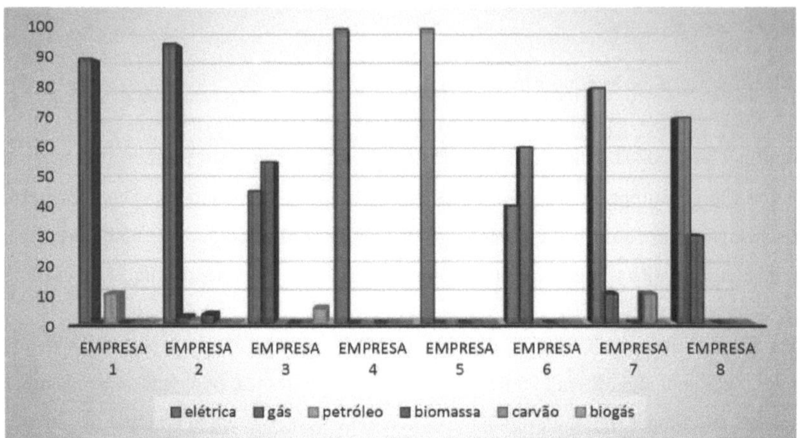

Graph 5 Use of Energy Sources.
Source: Prepared by the author.

Graph 5 shows that other sources account for a very low percentage, corresponding to approximately only 10% of use. However, two companies stand out for their use of natural gas as an alternative source, surpassing electricity with 55% and 60% of use.

Question 7, which sought to identify in which systems energy sources are most used within organisations, reported that 100% of energy sources are used in compressed air and lighting systems, followed by steam, refrigeration and motive power systems, which account for more than 70%, as shown in Graph 6 below.

Also noteworthy in this regard is the large share of motive power systems, in which several organisations have equipment and motors that cause high energy demand, due to the size of the machine and the length of time it is used, demonstrating the need for greater attention to be paid to this sector by companies (Graph 6).

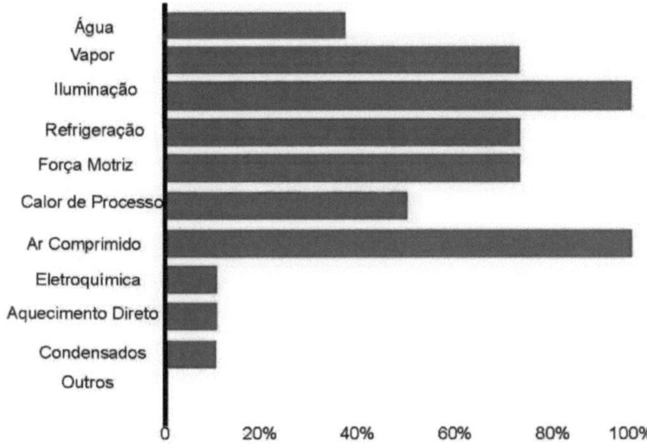

Graph 6 Systems Using Energy Sources.
Source: Prepared by the author.

In relation to the second descriptive question of the study (question 8), which sought to identify which equipment or processes consume the most energy, it can be seen that air compressors are the equipment that consume the most energy, followed by refrigeration and lighting systems, and finally steam generation and vulcanisation processes, as shown in Table 3.

Table 3 Equipment or Processes that Consume Energy.

COMPANY	EQUIPMENT OR PROCESSES
1	Manufacturing Process; Paper Impregnation Process; Melaminisation Process; Lighting; Administrative Use.
2	Air compressors.
3	Steam generator; Compressors.
4	Cooling system.
5	Air compressors; Lighting.
6	Vulcanisation; heating ovens; clothing.
7	Refrigeration.
8	Compressed air.

Source: Prepared by the author.

Table 3 shows that although this question is very specific to each industry duc to the types of equipment and areas of activity, air compressors were highlighted by three companies, while two emphasised refrigeration systems as the biggest consumer of electricity. On the other hand, two companies highlighted different processes and equipment.

With regard to the last question in this block (question 9), which sought to identify the companies' monthly MWH consumption, it was found that 50% of the industries classify their average monthly consumption between 0 and 400 MWH, followed by 25% between 401 and 600 MWH. In this case, only one company reported that its monthly consumption was between 801 and 1000 MWH, and only one above 1000 MWH (12.5%), as shown in Graph 7.

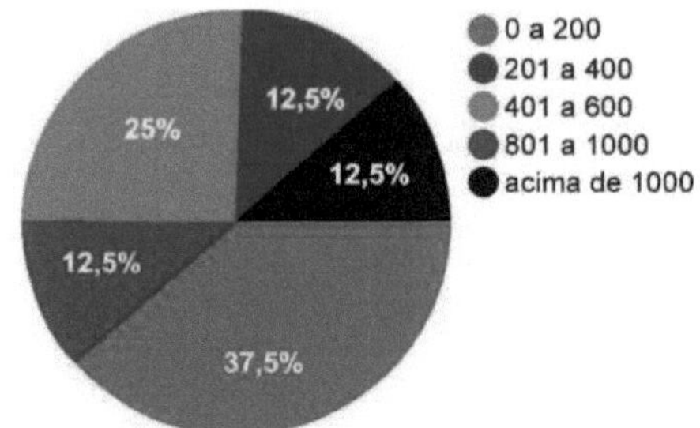

Graph 7 Monthly consumption in MWH.
Source: Prepared by the author.

4.1.3 Data and Results from the Third Block

The third and final block of questions sought to conclude the survey with complementary questions (10 - 17). This block sought to achieve the third specific objective, which is to identify the energy efficiency actions used in the organisations interviewed.

Question 10 sought to identify the objectives and targets set by organisations in relation to energy efficiency. Among the 9 options stipulated in the question according to the reference, it was found that approximately 90% of the objectives are related to saving energy in the most intensive processes, followed by investments in new strategies/technologies and reducing consumption by 0.5% or more, which accounted for more than 60% of the options in the questionnaire, and 50% of the targets are related to replacing equipment, as shown in Graph 8.

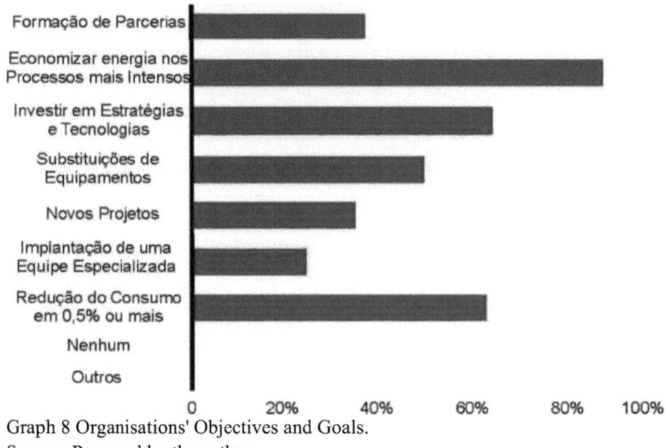

Graph 8 Organisations' Objectives and Goals.
Source: Prepared by the author.

Questions 11 and 12 sought to identify the existence of a CICE and whether companies are able to monitor and identify the processes and equipment that generate energy losses. It was found that 100% of the companies manage to control their processes and equipment that generate the greatest energy losses, which is a positive aspect as it demonstrates the organisation's governance and knowledge of energy losses in its processes and equipment.

With regard to the existence of a CICE within industries, it was found that half of the organisations do not have a team specialised in monitoring, identifying and implementing solutions in the area of energy efficiency (Graph 9). This question shows that 50 per cent of companies do not attach any importance to the creation of a CICE, which is fundamental for energy management systems.

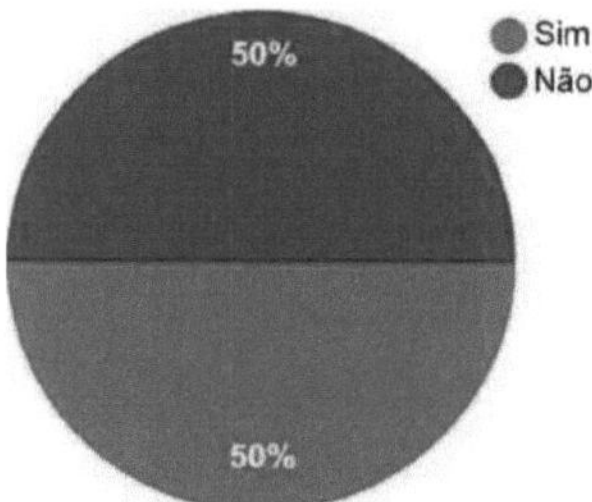

Graph 9 Companies with CICE.
Source: Prepared by the author.

Question 13ª sought to measure whether or not organisations carry out energy efficiency training for their employees, in which it was found that around 60% of companies carry out some kind of training or exercise related to energy efficiency for their employees, as shown in Graph 10. Still in this case, only 3 industries reported that they do not carry out any kind of energy efficiency training, which is a partially positive result.

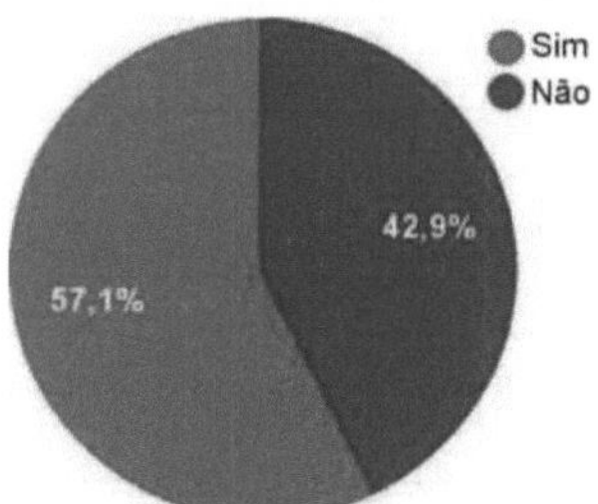

Graph 10 Energy efficiency training.
Source: Prepared by the author.

In order to assess whether industries measure energy consumption in the present and in the past, as well as whether they have a public/private partnership for energy management (questions 14 and 15), Graph 11 shows that more than 80% of organisations monitor electricity consumption in the present and in the past, which is a positive aspect as it demonstrates the concern of most companies with energy expenditure and their energy performance over time, thus being able to make consumption estimates and plan future improvements in this sector.

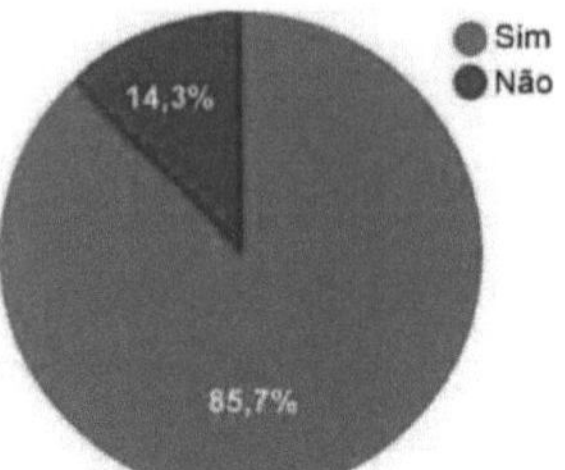

Graph 11 Evaluation of present and past consumption.

Source: Prepared by the author.

On the other hand, regarding the existence of a public/private partnership (question 15), it was found that only one of the companies surveyed has a partnership, and more than 80% of the sample said they did not have any kind of partnership aimed at energy efficiency programmes (Graph 12). This is a negative aspect for the region, since partnerships between the public and private sectors usually seek integration and planning focused on a specific issue that brings good results for both parties, in this case the implementation of energy efficiency programmes.

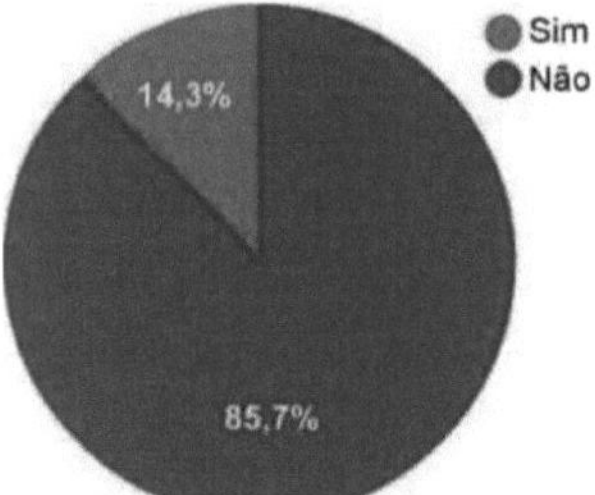

Graph 12 Public-Private Partnership.
Source: Prepared by the author.

Question 16 made it possible to check whether industries evaluate significant energy use when purchasing equipment. With regard to this, it was found that all the companies surveyed make these assessments before purchasing any type of equipment used within their processes and uses, totalling 100% of the sample, as shown in Graph 13.

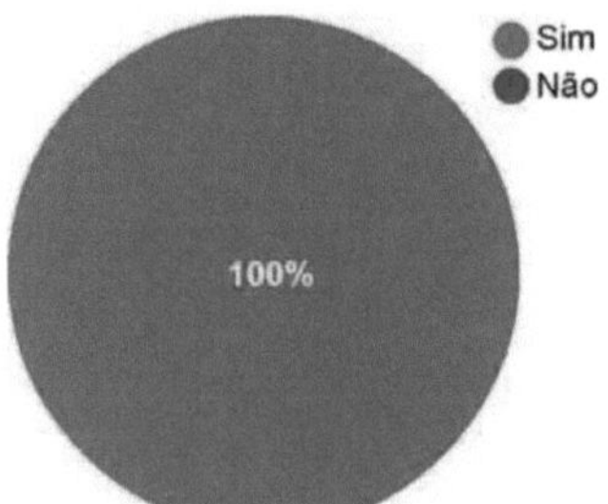

Graph 13 Evaluation of Energy Consumption in Equipment Purchases.

Source: Prepared by the author.

To finalise the questionnaire, question 17 sought to identify the types of assessments companies make when purchasing energy. In this case, all of them emphasised that the main item is cost versus benefit (100%), followed by energy quality with approximately 90% and availability with 75%, as can be seen in Graph 13.

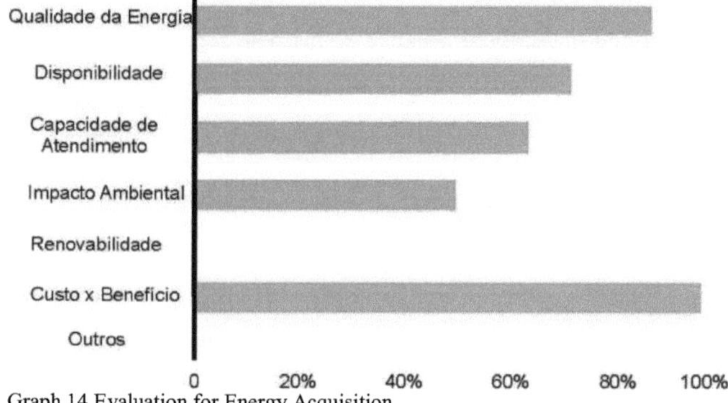

Graph 14 Evaluation for Energy Acquisition.
Source: Prepared by the author.

This question shows how the cost versus benefit, availability and quality of energy are the main criteria for evaluating a good purchase or acquisition of this resource. However, a very important aspect was not highlighted, such as the environmental impact, in which only 4 organisations said they assessed this item at the time of purchase and none of them reported measuring renewability. In a globalised world and with great worldwide concern about environmental issues, this aspect is likely to change in the future as this issue must be observed in all spheres and industrial activities, especially when it comes to purchasing a source of electricity.

CHAPTER 5

CONCLUSIONS

In accordance with the theoretical framework used and the correlation of the specific objectives addressed at the beginning of the study, this conclusion will present the final considerations of this research, the main points that stood out and the negative points with regard to the energy management of industries in the Ponta Grossa region, as well as suggestions for future work in the field of energy efficiency.

5.1 Final thoughts on the proposed objectives

In order to answer the general objective of the research, which is to investigate how industries in Ponta Grossa deal with the issue of energy management, the relevance of the topic in question was verified, in which through the bibliographical review cited by various authors it was argued that this issue has currently become extremely prominent, given that electricity today is one of the essential goods for humanity and is linked to the environmental issue through the preservation of natural resources.

On this subject, logistics also stands out, because through their processes and management of the flow of materials and services, energy generation and transmission companies can offer and make this resource available to companies with efficiency and quality. In addition, without electricity, many logistics processes are jeopardised, such as storage systems, production, billing processes and stock control, in other words, energy is fundamental to the functioning of logistics within a company.

Logistics also stands out through the transport of alternative energy sources, such as pipeline transport used to move natural gas and oil derivatives that are essential for energy generation and transmission. The important role of logistics in adapting to the new global trend towards the use of renewable energies can be seen, as a high level of efficiency in coordinating the various activities in the supply chain will lead to a reduction in operating costs related to energy and, above all, environmental preservation.

All these aspects confirm the importance of saving electricity, given that industries have the highest consumption of this resource in the country, as emphasised above.

Taking these aspects into consideration, the study's specific objectives were drawn up, which sought to emphasise how the industries of Ponta Grossa act on the issue in question. Through the first objective, which is to identify whether the companies have an energy management policy, it was observed that although the research was not aimed at an exclusive branch of activity, most of the organisations were consistent and aware of their energy management policy, reporting that they have an official declaration with written documents on the use, procedures and commitments to energy management matters. The only negative point in this block was the existence of an ISO 50001 certification for energy management systems, which none of the organisations currently has.

In the second specific objective on the sources of energy consumption in industry, it was also found that the majority of companies use other alternative sources of energy for their specific processes and systems, in which a large part of the energy is used in refrigeration, compressed air and lighting systems, which in this case is the particularity of each company because the use of energy depends on the type of equipment and processes in place. In this respect, it was also observed that the majority of companies depend on electricity, two of which said they were totally tied to this resource, as they had no alternative energy sources. The majority of organisations that use other sources have a very small share.

However, still in relation to renewable energy sources, it is understood that the energy generated by hydroelectric power stations is considered clean energy, as it has little environmental impact compared to non-renewable sources such as nuclear power and oil derivatives. However, it is necessary to diversify and use other sources in order to reduce dependence on conventional energy, reducing cogeneration and the risk of a blackout. However, this use depends a lot on availability and cost, and it was observed that none of the industries in the region use modern energy sources such as solar and wind power, perhaps due to the lack of government incentives and availability on the market.

In the last specific objective, which is to present the energy efficiency actions that organisations use, we found some positive points in relation to the fact that most of the companies carry out training for their employees and also all of them highlighted the fact that they monitor their energy losses, as well as purchasing equipment and evaluating consumption in the present and past. It was also noted that the biggest goal of the companies interviewed is related to saving energy in the most intensive processes, investing in new technologies/strategies and reducing electricity consumption. The negative points of this block are the lack of teams specialising in energy management in the industries, the precarious existence of public/private partnerships for energy management programmes and the lack of importance given to environmental issues when purchasing electricity.

The limitations of this research can be summarised as the industries that didn't answer the questionnaire due to their constant operations, which meant that they didn't have adequate time to analyse and explain the form. Another issue is the strategic nature of the companies' policy of maintaining the confidentiality of information and also the short duration of the case study.

In general, it can be concluded that the Ponta Grossa region has shown a good attitude towards energy management, despite the identification of some negative points in the survey, the majority of companies confirmed their concern and knowledge of the issue in question. This study has also provided information so that industries in other locations can realise the importance of managing this resource and adopt some of the practices described here, as well as promoting energy awareness.

5.2 Suggestions for future work

By carrying out this research, it was possible to verify some themes that could be developed as further scientific work in the field of energy management, such as:

- A broader study on energy management in industries with a larger population and sample, for example in industries in the Campos Gerais region or in the state of Paraná, in order to obtain even more significant results;

- To identify the costs and benefits for companies in the Ponta Grossa region related to the use of other renewable energy sources;
- A study to evaluate the cost/benefit advantages of the practices used by organisations in Ponta Grossa in other companies that do not have good energy management;
- Carrying out a study to assess the impacts that increased electricity consumption could have on the Ponta Grossa region in the future.

REFERENCES

BRAZILIAN ASSOCIATION OF TECHNICAL STANDARDS - (ABNT/CCE-116). ENERGY MANAGEMENT SYSTEM: Requirements with guidelines for use. Available at: <http://www3.fsa.br/localuser/energia/ABNT%20NBR%20...pdf>. Accessed on: 12 Sep 2015 at 16:25h.

INTERNATIONAL ENERGY AGENCY - IEA 2003. PETROL - ANEEL. Available at: <www.aneel.gov.br/aplicacoes/atlas/pdf/07-Petroleo(2).pdf>. Accessed on: 29 Aug 2015.

NATIONAL ELECTRICITY AGENCY - ANELL 2007. Disponível em: <http://www.mme.gov.br/documentos/10584/1432020/Matriz+Energ%C3%A9tica+Brasileira+2030+-+%28DOC%29/c220e91c-ab6d-4352-8605- 212888025a3a?version=1.0>. Accessed on: 05 May 2015.

ALEXANDRINI, F.et al. Study on the Brazilian Potential for Electricity Generation and the Creation of the Photovoltaic Industry - XXI SIMPEP, 2014. Available at: <www.simpep.feb.unesp.br/anais_simpep.php?e=9>. Accessed on: 27 June 2015.

ALVES, H. J.; MELCHIADES, F. G. and BOSCHI, A. O. Initial Survey of Thermal and Electrical Energy Consumption in the Brazilian Ceramic Tile Industry. Available at: <http://www.ceramicaindustrial.org.br/pdf/v12n01/v12n1a03.pdf>. Accessed on: 24 August 2015.

COMMERCIAL, INDUSTRIAL AND BUSINESS ASSOCIATION OF PONTA GROSSA. Competitive Ponta Grossa Programme. Available at

<http://www.acipg.org.br/pgcompetitiva/index.swf>. Accessed on: 16 Apr 2015 at 11:23h.

BRAZILIAN ELECTRIC ENERGY ATLAS: Conversion Factors - 2008. Available at: <www.aneel.gov.br/arquivos/PDF/atlas_fatoresdeconversao_indice.pdf>. Accessed on: 28 Apr 2015.

NATIONAL ENERGY BALANCE. Final Report 2013. Available at: <https://ben.epe.gov.br/downloads/Relatorio_Final_BEN_2013.pdf>. Accessed on: 15 Apr 2015.

NATIONAL ENERGY BALANCE. Synthesis Report 2014. Available at:< https://ben.epe.gov.br/downloads/Síntese do Relatório Final_2014_Web.pdf>. Accessed on: 16 Apr 2015.

BALLOU, R. H. Supply chain management. Business logistics; translated by Raul Rubenich. 5. ed. Porto Alegre: Bookman, 2006. 216 p.

CNI - NATIONAL CONFEDERATION OF INDUSTRY - 2014. Available at: <http://arquivos.portaldaindustria.com.br/app/conteudo_18/2014/04/22/6281/cartilha cni_corrente_FINAL-small1.pdf>. Accessed on: 08 Jun 2015.

COPEL - Manual for Energy Efficiency in Industry - 2005. Available at: <http://www.copel.com/hpcopel/root/sitearquivos2.nsf/arquivos/manual/$FILE/manual _eficiencia_energ.pdf>. Accessed on: 28 May 2015.

INTERNAL ENERGY CONSERVATION COMMISSION - CICE 2015. Available at: <http://www.cice.ufla.br/site/index.php>. Accessed on: 10 Jun 2015 at 16:30h.

DIARIO DOS CAMPOS. Diário dos Campos newspaper from Ponta Grossa: PG is 59ª in the country out of 200 with the most active companies. Available at: <http://www.diariodoscampos.com.br/economia/2015/04pg-e-59a-no-pais-entre- 200-with-more-active-companies/1384497/>. Accessed on: 16 Apr 2015 at 10:30.

DINIZ, C. R. and SILVA, I. B. SCIENTIFIC METHODOLOGY: Types of methods and their application. Available at: <http://www.ead.uepb.edu.br/ava/arquivos/cursos/geografia/metodologia_cientifica/ Met_Cie_A04_M_WEB_310708.pdf>. Accessed on: 06 Oct 2015 at 16:40h.

FARIA, M. A. F. Prospecting Methodology for Small Hydroelectric Power Plants. Available at: <http://www.aneel.gov.br/biblioteca/trabalhos/trabalhos/Dissertacao_Felipe_Faria.pdf>. Accessed on: 22 Apr 2015.

FERREIRA, C. A. et al. Performance of Eletrobrás, through Procel, in the Energy Efficiency of Brazilian Industries. Available at: <http://www.eletrobras.com/elb/data/documents/storedDocuments/%7BAEBE43DA- 69AD-4278-B9FC-41031DD07B52%7D/%7BE338A9B4-F851-4C4A-A853-6DA690C76B7A%7D/Atua%E7%E3o%20da%20Eletrobras%20atrav%E9s%20do%2 0Procel.pdf> Accessed on 21 Sep 2015 at 17:15.

FRAGOMENI, C. e GOELLNER, C. O Impacto no Meio Ambiente pela Atividade de Geração de Energia Elétrica pelo Uso de Recursos Hídricos. Available at: <http://www.upf.br/seer/index.php/rjd/article/view/2132>. Accessed on: 30 June 2015.

GIL, A. C. Como Elaborar Projetos de Pesquisa. 4. ed. São Paulo: Atlas, 2002. 176 p.

GIL, A. C. Como Elaborar Projetos de Pesquisa. 5. ed. São Paulo: Atlas, 2010.

GOLDEMBERG, J. and LUCON, O. Energy and the environment in Brazil. 2007. Available at: <http://www.scielo.br/pdf/ea/v21n59/a02v2159.pdf>. Accessed on: 12 March 2014.

GOLDEMBERG, J. Research and Development in the Energy Area. 2000. Available at: <www.scielo.br/scielo.php?script=sci_arttex&pid=S0102- 88392000000300014>. Accessed on: 22 Apr 2015.

GOMES, F. P. and TORTATO, U. Reverse Logistics Planning and Management in the Electricity Sector: a case study. Available at: <http://www.simpoi.fgvsp.br/arquivo/2010/artigos/E2010_T00241_PCN46506.pdf>. Accessed on: 21 Sep 2015 at 17:00.

JANNUZZI, G. M. Energy and the Environment. 2001. Available at: <pessoal.educacional.com.br/up/4660001/6249852/artigo_1_etapa_2_fisica_IGL__. pdf>. Accessed on: 18 Aug 2015.

LARANJEIRA, D. Transformations in the Energy Sector as a Competitive Advantage Factor for Logistics. Available at: <http://www.logisticamoderna.com/opiniao/890-as-transformacoes-do-setor- energetic-as-a-factor-of-competitive-advantage-for-logistics> Accessed on: 09 Sep 2015 at 14:05h.

LESTINGUI, M. D; MORENO, L. M. e SAVOIA, R. Energia Para Quê e Para Quem no Brasil. Available at: <http://www.boell-latinoamerica.org/downloads/energia09_port_Apresentacao_LAREF_2009_-_celio_bermann%20(1).pdf>. Accessed on: 24 Aug 2015.

LIMA, R. A. RENEWABLE ENERGY PRODUCTION AND THE SUSTAINABLE DEVELOPMENT: An analysis of the climate change scenario. 2012. Available at: <www.google.com.br/url?sa=t&rct=j&=&esrc=s&source=web$cd=1&ved=0CC0QFjAAahUKEwjEgee3z7fHAhWFipAKHb0rDu4&url=http%3A%2F%2Fwww.periodicos.ufrn.br%2Fdireitoenergia%2Farticle%2Fdownload%2F5145%2F4126&ei=fsLVVcTKCow=WVngS917jwDg&ugs=AFQjCNFq-ZY9ljceH0eWjmcm802ME- FXbw=bv.99804247,d.Y2I>. Accessed on: 20 Aug 2015.

MASCARENHAS, S. A. Scientific Methodology. Pearson Education do Brasil. São Paulo, 2012. 1.ed. 136 p.

MAYER, F. D; CASTELLANELLI, C. and HOFFMANN, R. ENERGY GENERATION FROM RICE HUSK: an environmental analysis. Available at: <http://www.abepro.org.br/bibliotecas/ENEGEP2007_TR650480_0007.pdf>. Accessed on: 06 Oct 2015 at 14:23h.

MANDIM, D. Statistics Uncomplicated. Vestcon. Brasília, 2004. 11ª ed. 229 p.

MINISTRY OF MINES AND ENERGY - Ten-Year Energy Expansion Plan - EPE 2014. Available at: www.epe.gov.br/Estudos/documentos/PDE2022.pdf. Accessed on: 22 Apr 2015.

MINISTRY OF THE ENVIRONMENT - ENERGY: The order is to economise. Available at: www.mma.gov.br/estruturas/secex_consumo/_arquivos/7 - mcs_energia.pdf. Accessed on: 03 Aug 2015.

NASCIMENTO, T. C. et al. R&D STRATEGIES AND SUSTAINABILITY IN THE ELECTRIC SECTOR: The case of an energy company. 2013. Available at: <www.spell.org.br/periodicos>. Accessed on: 18 Aug 2015.

PACHECO, F. RENEWABLE ENERGIES: Brief concepts. Available at: <scholar.google.com.br/scholar?q=energias+renováveis&btnG=&hl=en- BR&as_adt=0%2C5>. Accessed on: 20 Aug 2015.

NATIONAL ENERGY EFFICIENCY PLAN. Basic Premises and Guidelines. Available at: <http://www.orcamentofederal.gov.br/projeto-esplanada- sustentavel/pasta-para-arquivar-dados-do-pes/Plano_Nacional_de_Eficiencia_Energetica.pdf>. Accessed on: 02 Sep 2015 at 21:39.

PORTAL BRASIL. INFRAESTRUTURA: Understanding how electricity reaches your home. Available at: <http://www.brasil.gov.br/infraestrutura/2014/08/entenda-como-a-energia-eletrica- chega-a-sua-casa>. Accessed on: 16 Sep 2015 at 11:33am.

PROCEL - National Electricity Conservation Programme. International Experiences in Energy Efficiency in Industry. Available at: <http://arquivos.portaldaindustria.com.br/app/conteudo_24/2012/09/06/262/20121127181849323774u.pdf>. Accessed on: 27 August 2015.

PROCEL. CASOS DE SUCESSO: Eficiência Energética na Vincunha Têxtil Unit III. Available at: <http://www.procelinfo.com.br/main.asp?TeamID={4CE15A4A-BB17-4A70-9148-C3A67E4DEBE2}>. Accessed on: 15 Oct 2015 at 16:31h.

REIS, L. B; FADIGAS, E. A. A e CARVALHO, C. E. Energia, Recursos Naturais e a Prática do Desenvolvimento Sustentável. Barueri, SP: Monoele, 2005. 1[a] ed. 415 p.

SOLA, H. V. A. and KOVALESKI, L. J. ENERGY EFFICIENCY IN INDUSTRIES: Scenarios and Opportunities. Available at: <pessoal.utfpr.edu.br/luizpepplow/arquivos/81.pdf>. Accessed on: 22 Apr 2015.

APPENDIX

Appendix A - Survey Instrument

QUESTIONNAIRE ON THE LEVEL OF ENERGY MANAGEMENT IN INDUSTRIES IN THE PONTA GROSSA REGION		
OBJECTIVE	**QUESTIONS**	**AUTHOR/SOURCE**
Identify whether companies have an energy management policy	1- what is the company's line of business? ( )food ( )chemical ( )petrochemical ( )metallurgical ( )automotive ( )industry ( )electronics ( )plastics ( )wood ( )agro-industry ( )other	- PNEE (2011).
	2- Is there an official company declaration with energy management commitments? ( )yes ( )no	- ABNT ISO 50001.
	3- Does the organisation have a certification for energy management systems (e.g. ISO 50001)? ( )yes ( )no	
	4- are there written procedures for the use of electricity? ( )yes ( )no	

Present the sources of energy consumption in industries	5- What renewable and non-renewable energy sources does the company use? ( )solar ( )wind ( )coal ( )biomass ( )hydro ( )gas ( )oil ( )uranium ( )geothermal ( )electric ( )other	- PACHECO (2006). - HINRICHIS et al. (2011) apud ALEXANDRINI et al. (2014).
	6- What is the percentage utilisation of each energy source? R:	- PNEE (2011). - FERREIRA et al. (2009).
	7- In the company, which energy sources are used in which systems? ( )water ( )steam ( )lighting ( )refrigeration (motive power ( )process heat ( )compressed air ( )electrochemical ( )direct heating ( )condensates ( )other	- PNEE (2011).
	8- what equipment or processes consume the most energy? R:	
	9- What is the average number of megawatt-hours consumed each month? ( )0 to 200 ( )201 to 400 ( )401 to 600 ( )601 to 800 ( )801 to 1000 ( )over 1000.	- ATLAS (2008).
Present the energy efficiency actions that organisations use	10- what are the company's objectives and targets for reducing energy consumption? (formation of public-private partnerships ( )Equipment replacements ( )invest in strategies and new technologies ( )reduce consumption by 0.5% or more ( )Saving energy in the most intensive processes ( )new projects ( )deployment of a specialised team ( )none ( )other	- PNEE (2011).
	11- is it possible to monitor equipment or processes that generate energy losses? ( )yes ( )no	- Sola and Kovaleski (2004).
	12- Is there a team responsible for monitoring, identifying and implementing energy efficiency solutions (e.g. Internal Energy Conservation Commission - CICE) ( )yes ( )no	- CICE (2015).
	13- Does the company provide energy efficiency training for its employees? ( )yes ( )no	
	14- Does the organisation carry out assessments of past and present energy use? ( )yes ( )no	ABNT ISO 50001. - PROCEL (2012). - PNEf (2011). - COPEL (2005).
	15- Does the organisation have a public/private partnership for energy efficiency programmes? ( )yes ( )no	
	16- When buying equipment, does the company assess its significant energy use before purchasing it? ( )yes ( )no	
	17- when buying energy, what does the company evaluate? ( ) power quality ( )availability ( )service capacity ( )environmental impact ( )renewability ( )cost-benefit ( )other	- ABNT ISO 50001.

Printed by Books on Demand GmbH, Norderstedt / Germany